Gustavo Berger Moura

Coating produced by plasma electrolytic oxidation on aluminum

Gustavo Berger Moura

Coating produced by plasma electrolytic oxidation on aluminum

Electrical properties

ScienciaScripts

Imprint

Any brand names and product names mentioned in this book are subject to trademark, brand or patent protection and are trademarks or registered trademarks of their respective holders. The use of brand names, product names, common names, trade names, product descriptions etc. even without a particular marking in this work is in no way to be construed to mean that such names may be regarded as unrestricted in respect of trademark and brand protection legislation and could thus be used by anyone.

Cover image: www.ingimage.com

This book is a translation from the original published under ISBN 978-3-330-76954-0.

Publisher:
Sciencia Scripts
is a trademark of
Dodo Books Indian Ocean Ltd. and OmniScriptum S.R.L publishing group

120 High Road, East Finchley, London, N2 9ED, United Kingdom
Str. Armeneasca 28/1, office 1, Chisinau MD-2012, Republic of Moldova, Europe
Managing Directors: Ieva Konstantinova, Victoria Ursu
info@omniscriptum.com

Printed at: see last page
ISBN: 978-620-8-62419-4

SUMMARY

I dedicate this work to all my family and friends!

ACKNOWLEDGMENTS

Even if I had the most brilliant mind to search for words of gratitude, they would not be enough to express my feeling of gratitude towards my GOD, but it is HIM that I thank first in this sincere list of thanks, thank you LORD GOD OUR CREATOR for giving me this great VICTORY!

I am very grateful to my advisor, Prof. Dr. Nilson Cristino da Cruz, who has given me the opportunity to carry out this work. I am grateful for the knowledge, experience, trust and patience he has passed on to me, which are fundamental requirements for carrying out this work.

To Prof. Dr. Elidiane Cipriano Rangel for her example of dedication and efficiency.

To Prof. Dr. José Roberto Bortoleto for his friendship, support and knowledge.

To Prof. Dr. Steven Durrant for the lessons learned.

To Prof. Dr. Sandro Mancini for the teaching I received with dedication.

Prof. Dr. Maria Lùcia Pereira Antunes "Malu" for always being willing to help.

To Prof. Dr. Francisco Trivinho Strixino for taking part in my exam and enriching this work.

To my friend Flâvio de Souza for showing me that this path was possible.

To my friend Brother Mauro Rube de Morais for his Christian friendship.

My wife Ana Maria, who encouraged me to carry on when I had given up.

To my children Matheus, Nathàlia and Luiz Gustavo who are GOD's gifts in my life.

To my parents Aureo Moura Jùnior and especially to my mother Jussara Berger for having taught me the right way.

To my brothers Igor and Rodrigo, and especially to my brother Vinicius, for the electrical knowledge they passed on.

To Dr. César Augusto Antônio for his support, patience, encouragement and light during my first steps in this work.

To Rafael Parra who, with great humility and dedication, helped me a lot in all the analyses at LaPTec.

My aunts Dalila Berger Arantes and Fabiana Berger Cerioni, who helped me through life with great love.

To my grandfather Paulo Berger for his example of life.

"I am the Alpha and the Omega, the beginning and the end, the first and the last".

Revelation, chapter 22, verse 13.

SUMMARY

In this work, plasma-assisted electrolytic oxidation (PEO) was used to produce a ceramic coating on the surface of aluminum alloy (AA 5052) substrates. The electrical properties of this coating were analyzed using electrical impedance spectroscopy (EIE). The thicknesses of the coatings were determined using the eddy current method and scanning electron microscopy (SEM). SEM was also used to evaluate the topography of the coatings. Infrared absorption spectroscopy (IRS) and energy dispersive x-ray spectroscopy (EDS) were used to determine the structure and chemical composition. The crystal structures were determined using the x-ray diffraction technique. The results revealed that the surfaces were coated with a complex coating, mainly containing aluminum, oxygen and silicon, which increased the electrical resistivity 10^{10} times compared to the as-received aluminum.

Keywords: Plasma-assisted electrolytic oxidation, PEO; Electrical properties of aluminum; Electrical resistivity; Spectroscopy

1 INTRODUCTION

Various surface treatments have been used to confer properties not found in the original material (ASM, 1992).

In particular, painting and anodizing processes involving harmful and highly toxic agents such as acids and chromates are often used to treat metals (DEHNAVI, 2013). The technique (PEO) proposed in this work, as it uses dilute alkaline electrolytes, does not harm the environment (GUPTA, 2007; JIANG, 2011).

The PEO process used in this work operates with a high-voltage pulsed current source, in which the positive electrode is connected to the sample and the negative electrode to the reactor containing an electrolytic solution. This process produces unique coatings, which are characteristic of this technique (YEROKHIN, 2005; GUPTA, 2007). On metals such as aluminum, these coatings offer greater protection against wear, corrosion (ANTÔNIO, 2011; OLIVEIRA, 2009) and electrical resistance, as studied in this work.

Electrical insulation on the surface of a metal such as aluminum is often desired in order to protect the system or circuit in which it is inserted as an electrical conductor.

The PEO-assisted aluminum samples had a ceramic coating formed on their surfaces. This coating had its electrical and chemical properties investigated in this work.

The parameters used in the PEO process were made feasible in this work in order to obtain the lowest possible electricity consumption, with a view to a possible industrial application, since electricity consumption is a preponderant factor in industrial production.

2 LITERATURE REVIEW

2.1 Plasma assisted electrolytic oxidation (PEO)

Plasma electrolytic oxidation (PEO) goes beyond conventional anodizing, producing ceramic coatings with plasma at atmospheric pressure (GUPTA, 2007).

In PEO processing, coatings are produced on metal substrates immersed in a slightly alkaline electrolyte, connected to the positive pole of the source, where the potential difference can reach 1000 V. At the start of processing, when the potential difference is small, a passive oxide film forms on the substrate, similar to anodizing. As the voltage increases, exceeding the metal's corrosion potential, a new, more porous coating is formed, with the first coating, which is now partly dissolved, as the precursor. When the dielectric strength of the oxide is broken by the intensity of the electric field, electric sparks are generated around the surface of the film, contributing to its growth (MONTERO et al, 1987). These sparks are transformed into electrical micro-arcs when the voltages are higher, which can raise the temperature to thousands of degrees Celsius in this region (DUNLEAVY, 2009);

MARTIN, 2013). This high heat melts the oxide layer, reacting with the electrolyte. This results in a complex coating made up of elements from the electrolyte, substrate and oxide layer. The properties of these films generally have high electrical, chemical and mechanical resistance (WEI, 2005; TIAN, 2002).

2.2 PEO x Anodizing

The process of anodizing aluminium is carried out at low voltages (~15 V), using a dilute acid (usually sulphuric acid) as an electrolyte, which, when dissociated in an aqueous solution, provides oxygen for the oxidation of aluminium without reacting with the oxide formed, favoring its growth (ASSOCIAÇÃO BRASILEIRA DO ALUMiNIO, 2004; LUGOVSKOY, 2013).

On the other hand, the processing of aluminum by PEO takes place at high voltages (greater than >100 V), with dilute alkaline electrolytes, which provide hydroxyls (OH^-) for the reaction:

$$Al_2O_{3(s)} + 3H_2O_{(aq)} + 2OH^-_{(aq)} \rightarrow 2Al(OH)_{3(aq)} + 2OH^-_{(g)} \quad (1)$$

Unlike conventional anodizing, which only oxidizes the metal, PEO-assisted film growth occurs through the action of micro-arcs at the substrate-electrolyte interface, producing a complex coating with elements from the electrolyte and substrate (LUGOVSKOY and BOSTA, 2013).

2.3 Characterization techniques

2.3.1 Thickness measurements using the eddy current method

Eddy currents are generated when a magnetic field variable induces electric currents in a conductive part.

Determining the thickness of anodic or organic layers is standardized by ABNT

NBR 12610:2010, which specifies the eddy current method for measuring the thickness of non-conductive layers in samples of aluminium and its alloys. The equipment used has one or more excitation coils which, when run by alternating current, produce a magnetic field in their vicinity. This field is used to induce eddy currents in the sample. These, in turn, induce secondary magnetic fields and eddy currents in the excitation coil. In this way, changes in the characteristics of the sample, such as the formation of an insulating surface layer, produce changes in the current flowing through the excitation coil which are detected and converted into a thickness measurement (HALLIDAY, 2009).

This technique can measure layers up to 5000 µm thick in non-ferrous metals, with a resolution of up to 0.1.

2.3.2 Scanning Electron Microscopy (SEM) and Energy Dispersive Spectroscopy (EDS)

The scanning electron microscope (SEM), because it uses an electron beam and therefore its resolution is not limited by the wavelength of visible light as in optical microscopes, allows magnifications of up to 1,000,000 times.

The electron beam generated by heating a filament is directed at the sample by applying an electrical potential difference. Electrostatic lenses and electromagnetic coils are used to collimate and move the electron beam over the sample.

When an electron beam hits a sample, various interactions can occur, such as

Auger electron emission, secondary electron emission, backscattered electrons, characteristic x-ray emission and x-ray fluorescence (SKOOG, 2002).

By detecting and processing this information, it is possible to analyze the elemental composition of the sample by measuring the energy of the emitted x-rays (EDS) and also to obtain images with compositional contrast from the backscattered electrons.

2.3.3 Infrared Absorption Spectroscopy

Although radiation in the infrared region of the electromagnetic spectrum does not have enough energy to excite electronic levels, it can be absorbed by a material, exciting its vibrational levels. In this way, by incident radiation with varying energy and measuring the energies that are absorbed, it is possible to identify the chemical bonds present in a given sample (ATKINS, 2001; SKOOG, 2002).

2.3.4 X-ray diffraction (XRD)

The x-ray diffraction (XRD) technique consists of emitting a monochromatic x-ray beam onto the sample and measuring the diffraction angle of the diffracted beam, which is proportional to the characteristic interplanar spacing of the crystal structure to be analyzed. For diffraction to occur, Bragg's Law must be met, which states that the parallel beams of incident radiation after diffraction interfere constructively according to the following equation:

$$n\lambda = 2d_{hkl} \text{ sen } \theta \quad (2)$$

Where n is the order of reflection, λ is the monochromatic wavelength and θ is the diffraction angle (CALLISTER, 2007; PADILHA, 2000).

2.3.5 Electrical impedance spectroscopy

Electrical impedance spectroscopy is a technique that uses electrical stimuli (voltage and/or current) to obtain an electrical response from a material (MACDONALD, 2005). This response is characterized by the impedance Z of the sample, which is made up of the resistance R and the capacitive reactances XC and XL. When these reactances are associated in series, the resulting impedance is given by:

$$Z = \sqrt{R^2 + (X_L - X_C)^2} \quad (3)$$

On the other hand, parallel elements are governed by the following expression:

$$\left(\frac{1}{Z}\right)^2 = \left(\frac{1}{R}\right)^2 + \left(\frac{1}{X_L} - \frac{1}{X_C}\right)^2 \quad (4)$$

The reactances XC and XL vary with frequency according to the following equations:

$$X_C = \frac{1}{\omega C} \quad (5)$$

$$X_L = \omega L \quad (6)$$

where C is capacitance in faraday, L is inductance in henry and

$$\omega = 2f\pi \tag{7}$$

where *f is the* frequency in hertz (CHINAGLIA, 2008).

3 EXPERIMENTAL PROCEDURE

The research was carried out at the Plasma Technology Laboratory (LaPTec) at UNESP's Experimental Campus in Sorocaba, as were the sample treatments and infrared, x-ray diffraction, scanning electron microscopy and x-ray energy dispersive analysis.

3.1 Sample preparation

The AA 5052 aluminium alloy samples were prepared in the dimensions 31 mm x 27 mm x 1.5 mm with a 20 mm x 7 mm appendage, as shown in Figure 1, to be connected to the treatment cell.

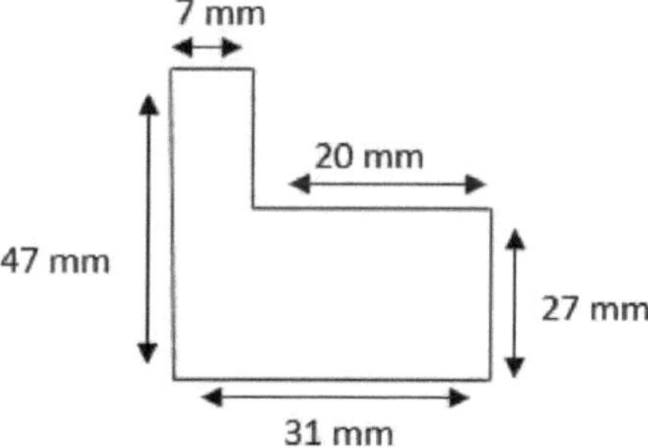

Figure 1: Sample dimensions.

The samples were prepared according to the flowchart shown in Figure 2.

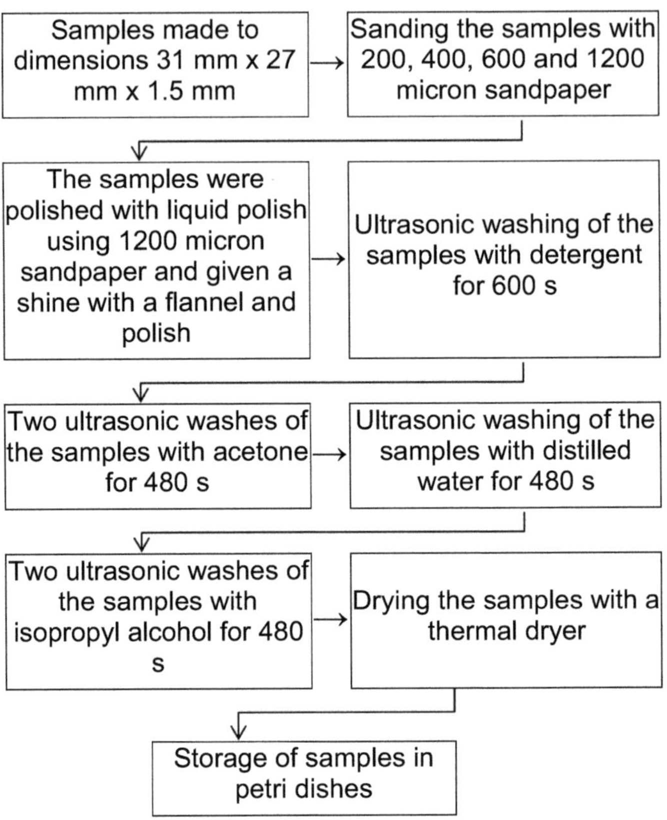

Figure 2: Flowchart of sample preparation steps

3.2 Preparation of the electrolyte solution

The solution used was prepared by dissolving 20 grams of sodium silicate (Na_2SiO_3) per liter of deionized and distilled water. This solution was chosen because it had proved to be effective in previous studies (YASUDA, 1985; YEROKHIN, 1999).

3.3 Sample treatment

A pulsed voltage source (Plasma Technology Ltd MAO-30) was used to treat the samples. The grounded terminal of this source is connected to a stainless steel vat containing the electrolyte solution (Figure 3). The other terminal is connected to a support in contact with the sample.

Figure 3: Photo of the electrolytic oxidation system, (1) sample holder connected to the positive terminal of the source, (2) stirrer, (3) sample, (4) stainless steel vat, (5) grounded terminal of the source and (6) water circulation tubes of the cooling system.

The samples were subjected to treatments lasting 60, 90, 120, 180, 360 and 540 seconds in duplicate, so that it would be possible to relate the characteristics of the coating to this parameter. The frequency used for the polarization pulses of the samples was 900 Hz, which has shown a good growth rate in previous studies (MARTIN, 2013). The current density was kept at 0.027 A/cm^2, which is the lowest possible for the source used in order to minimize energy consumption. With this current density, the maximum voltage reached 375 V after 540 s of treatment. A digital thermometer was used to monitor the electrolyte temperature, which varied from 29.0 °C (after 60 s) to 31.8 °C (for 540 s treatments).

3.4 Coating characterization techniques

3.4.1 Thickness by eddy current method

The device used was the Automation Dr. Nix Model Qnix 7500 (Figure 4). To carry out the measurement, the samples were placed on a smooth, clean surface lined with paper towels. The measurement was made by placing the probe perpendicularly in contact with the surface of the sample at 6 different points on the surface of the coating, avoiding the edges.

Figure 4: Photo of the non-conductive layer thickness meter using eddy currents . Automation Dr. Nix Model Qnix 7500.

3.4.2 Scanning electron microscopy and energy dispersive x-ray spectroscopy

To analyze the topography and thickness by SEM and the chemical composition of specific points of the sample by EDS, the JEOL JSM-6010 electron microscope from LaPTec at UNESP Sorocaba was used. This equipment is capable of analyzing elements with atomic numbers ranging from beryllium to uranium, with a resolution of 129 to 133 eV.

Topographic analysis was carried out using micrographs obtained with secondary electrons accelerated by a voltage of 3 kV.

A beam with an acceleration voltage of 10 kV was used to analyze the chemical composition.

The thickness of the coating was measured on the cross-section of the sample at three different points with 10 measurements at each point. These measurements were made using a backscattered electron beam from primary electrons with an accelerating voltage of 10 kV. The image was obtained from an effective area of 10 mm2.

3.4.3 Infrared absorption spectroscopy

The spectrometer used was a Jasco FTIR-410 from LaPTec at UNESP Sorocaba. The samples were analyzed using the reflectance-absorbance technique (IRRAS) based on the co-addition of 128 spectra in the range of 400 to 4000 cm^{-1} with a resolution of 4 cm^{-1}.

3.4.4 X-ray diffraction (XRD)

A PANalytical X'Pert PRO dilatometer from LaPTec at UNESP Sorocaba was used to determine the crystalline phases present in the coatings.

The analyses were carried out using Cu Ka radiation with a wavelength of 0.1540 nm, an accelerating voltage of 45 kV, a current of 40 mA, a beam with an angle of incidence of 1°, a step of 0.05° and 2 s per step.

3.4.5 Electrical impedance spectroscopy

The electrical properties of the coatings were investigated using a SOLARTRON Analytical impedance spectrometer installed at LaPTec at UNESP Sorocaba (Figure 5).

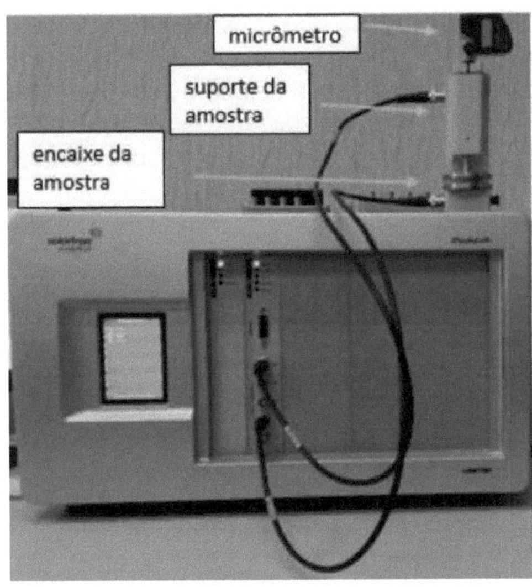

Figure 5: Photograph of the SOLARTRON Analytical electrical impedance spectrometer.

The samples are positioned between two electrodes whose separation, controlled by a micrometer screw, is indicated by a digital display. In all cases, the voltage applied was 5 V and the frequency ranged from 10 to 100,000 Hz.

4 RESULTS AND DISCUSSION

4.1 Definition of sample processing conditions

No studies were found in the literature characterizing the electrical properties of coatings formed on aluminium alloys by PEO using the electrical impedance spectroscopy (EIE) technique. In this way, the process parameters were selected with the aim of achieving not only good coating properties, but also lower electricity consumption for possible future industrial application.

The frequency chosen for the treatments was 900 Hz because, according to reports in the literature (YEROKHIN, 2005; KHAN, 2010), this frequency resulted in good growth rates.

Once the frequency and current density were fixed, the other process parameters varied as shown in Table 1. The increase in voltage so that the current density remained constant with the treatment time is a consequence of the increase in the thickness of the insulating layer formed on the substrate. This increase in voltage leads to an increase in the power dissipated by the system source, leading to the observed increase in electrolyte temperature.

Table 1: Sample treatment parameters using PEO. In all cases, the current density was kept constant at 27 mA/cm^2

Time (^{s})	Voltage (V)	Temperature (°C)
60	190	27,3
90	321	28,6
120	343	28,1
180	346	30,8
360	370	30,4
540	375	31,8

4.2 Coating thicknesses

The thicknesses of the coatings formed when the samples were treated with PEO were measured using eddy current and SEM methods.

Figure 6 shows a scanning electron micrograph of the cross section of a sample treated for 540 s where the thickness values are shown in dots.

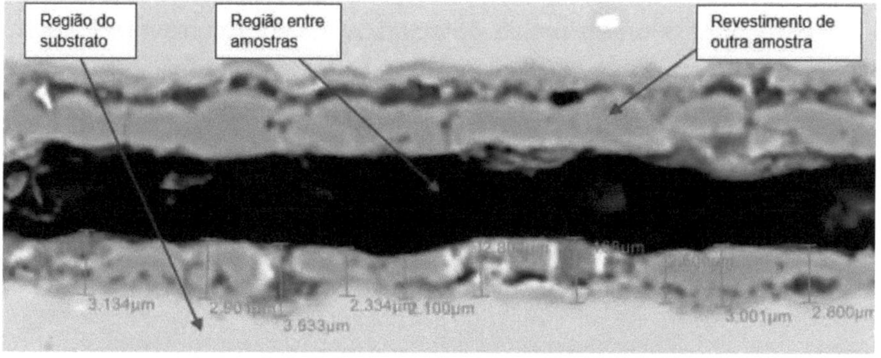

Figure 6: Image of the cross-section of a sample treated for 540 s showing the values obtained for the coating thickness in 10 positions.

Figure 7 shows the average values of the coating thicknesses analyzed by the two techniques as a function of treatment time.

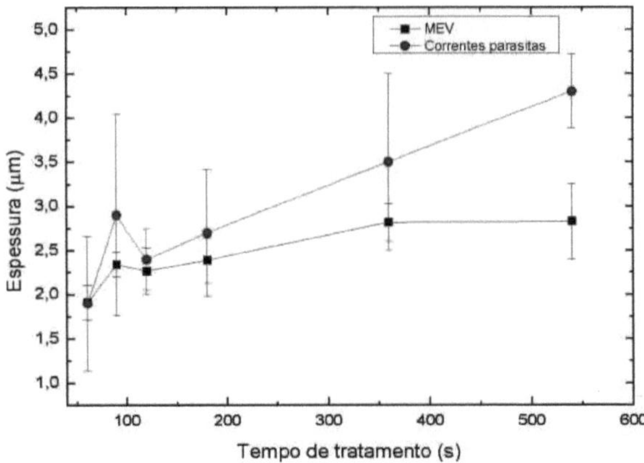

Figure 7: Coating thickness as a function of treatment time determined by SEM and eddy currents.

The measurements made using the eddy current technique showed greater thicknesses, with greater standard deviations, compared to the results obtained using SEM. These higher values can be attributed to the greater influence of roughness on the results obtained by the eddy current technique.

Determining the thickness of coatings by SEM has the advantage that the thicknesses are measured in microscopic regions, which allows for many measurements with little variation due to the roughness of the sample, which is confirmed by the low standard deviation values seen in Figure 7.

In the first 90 s of treatment, a relatively high growth rate was observed (over 1 µm/min) due to the strong interaction of the PEO with the low electrical resistance metal substrate. Thus, with the growth of the coating, the increase in the electrical resistance of the substrate leads to a decrease in the growth rate for longer treatments.

For the calculations of the coating's electrical properties, which require thickness, the thicknesses determined by SEM were taken into account, as this provided the greatest number of measurements and the lowest standard deviation.

4.3 Morphology and elementary composition

Scanning electron microscopy (SEM) was used to study the morphology of the sample surfaces. In this analysis, SEM uses secondary electrons which have low energy, thus providing topographical information of the surface. Figures 8 and 9 show the micrographs of the surfaces of the samples, untreated and those treated with PEO for various treatment times respectively. As can be seen in Figures 8 and 9, as the treatment time increases, the number and size of pores increases.

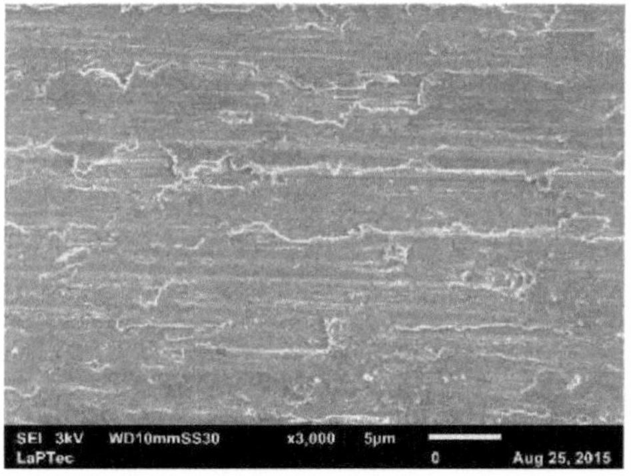

Figura 8: SEM micrograph of the untreated sample

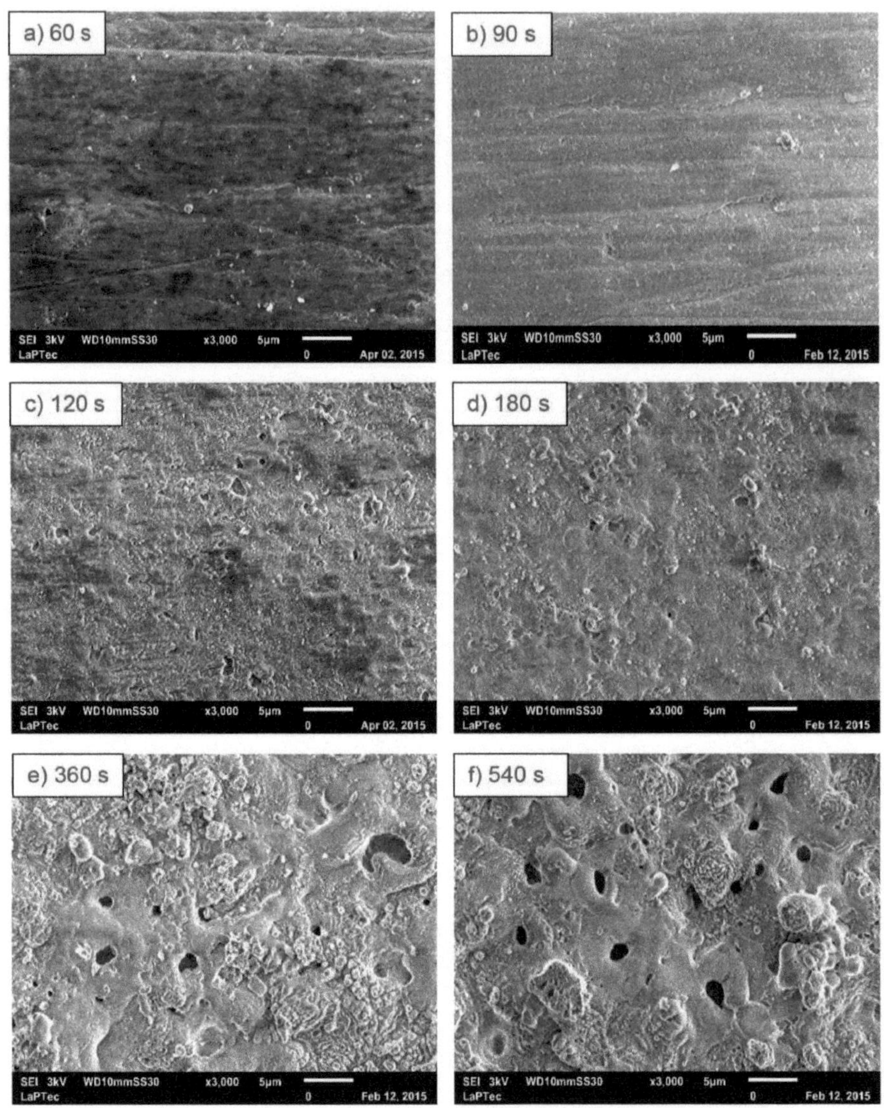

Figura 9: SEM micrographs of samples treated with PEO at various times: (a) 60 s, (b) 90 s, (c) 120 s, (d) 180 s, (e) 360 s and (f) 540 s.

Figure 10 shows the variation in the average concentrations of aluminum, silicon and oxygen, obtained by EDS, as a function of treatment time. As can be seen, as the treatment time increases, the incorporation of oxygen and silicon from the electrolyte into the substrate increases. As a result, the originally metallic surface becomes a layer composed predominantly of aluminum oxide and silicon.

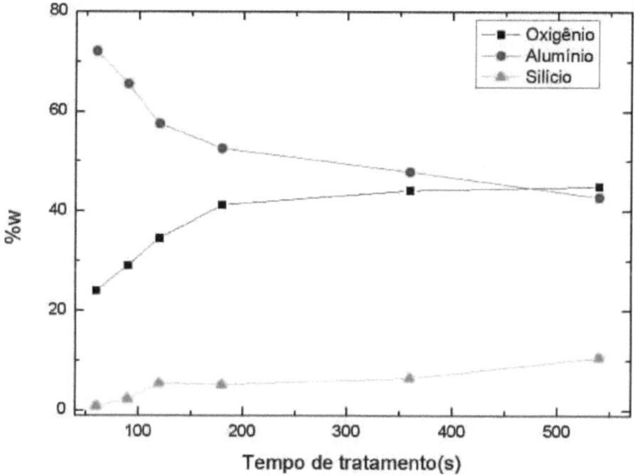

Figure 10: Concentration of aluminum, oxygen and silicon in samples with different PEO treatment times.

Although the electrolyte is composed of sodium silicate, there was no incorporation of Na into the coatings. One reason for this could be that the Na cation^{+} is attracted to the negative pole of the source (cathode), the opposite side to the substrate.

4.4 Chemical composition by absorption spectroscopy in the infrared (FTIR)

Figure 11 shows the infrared absorption spectra for the samples that were subjected to treatments lasting from 60 to 540 s.

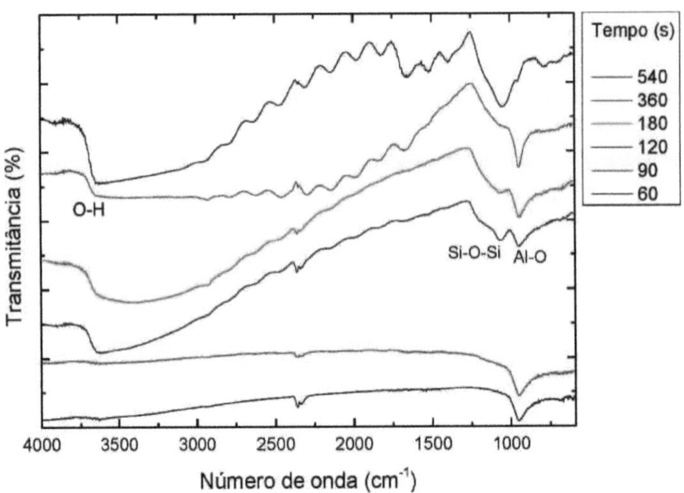

Figure 11: Infrared absorption transmittance spectra of the coatings produced on aluminum by PEO at different treatment times.

From 120 s of treatment with PEO, an absorption band is observed at 3650 cm^{-1} which increases until 540 s of treatment. This band is attributed to the stretching of the hydroxyl group (O-H). This bond probably originates from the aqueous solution used in the treatment and/or the adsorption of moisture on the surface of the coating (LIF, 2007; BRAND, 2004).

As the treatment time increased, the absorption band related to the stretching of the Si-O-Si bond was observed at approximately 1150 cm^{-1}. The incorporation of silicon into the structure of the coating is attributed to the fact that a Na2SiO3 solution was used as the electrolyte. When in aqueous solution, Na2SiO3 disassociates into 2Na^{+} and SiO_3^{2-} and SiO_3^{2-} is attracted to the positive pole of the source, reacting with the coating (WEI-CHAO, 2011).

At 970 cm^{-1}, another absorption band was observed, the intensity of which gradually increased with treatment time. Haanappel related the absorption in this band to the Al-O bonds in aluminium oxide (HAANAPPEL, 1995).

4.5 Crystalline structure of coatings

Figure 12 shows the diffractograms obtained with the samples treated for 60, 90, 120, 180, 360 and 540 s.

As can be seen, only crystalline phases of metallic aluminium (PDF 85-1327) and aluminium oxide were obtained. The intensities of the aluminium oxide gamma phase peaks (PDF 79-1557), which only appeared in samples treated for more than 180 s, became more intense with increasing treatment time. Although the presence of silicon in the coatings was verified by EDS and FTIR, no crystalline structures containing this element were observed. This fact, which has also been observed by other authors (LIF, 2007; LISBÔA and LI 2014), indicates the incorporation of silicon into the coating in the form of amorphous silicon oxide.

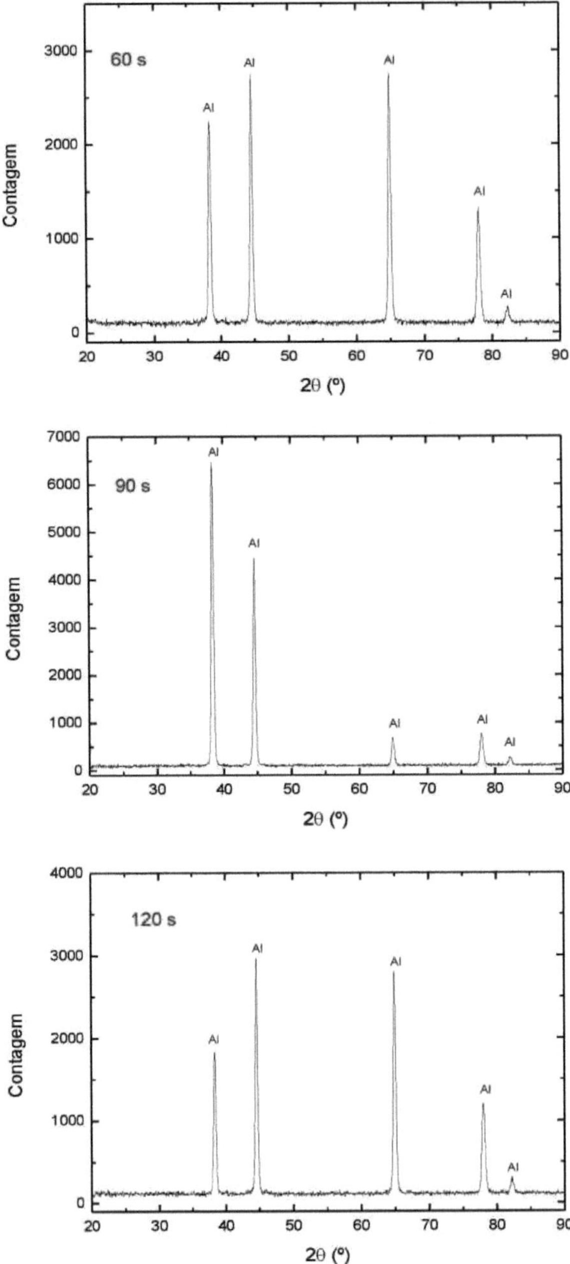

60 s
Al
Al
Al
Al
Al
Contagem
3000
2000
1000
0
20
30
40
50
60
70
80
90
2θ (º)
90 s
Al
Al
Al
Al
Al
7000
6000
5000
4000
3000
2000
1000
0
Contagem
2θ (º)
120 s
Al
Al
Al
Al
Al
4000
3000
2000
1000
0
Contagem
2θ (º)

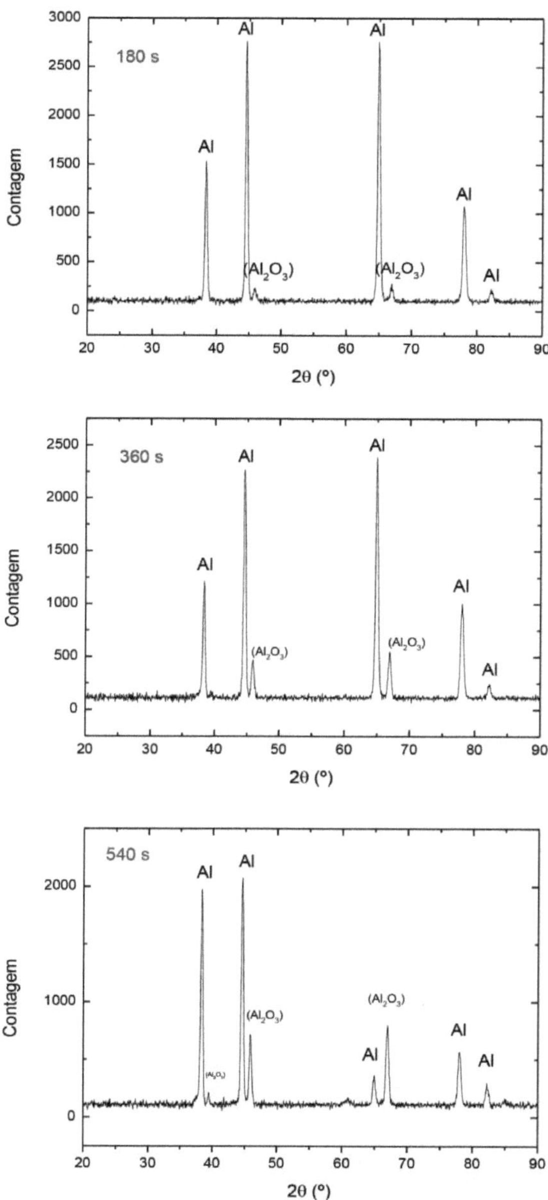

Figure 12: X-ray diffractograms of the samples treated at various times.

4.6 Electrical properties of coatings

Figure 13 shows an equivalent circuit used to simulate and calculate the electrical components of the coatings.

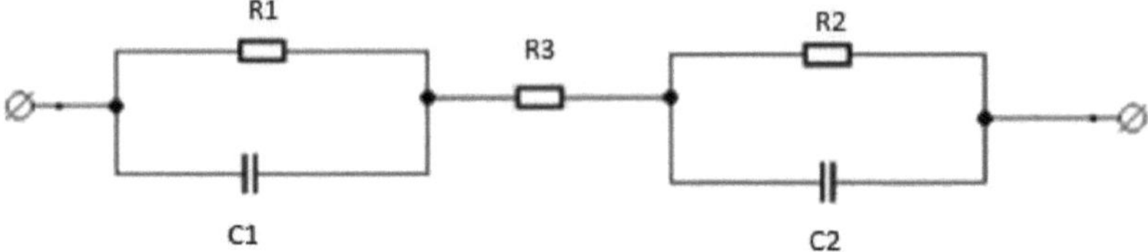

Figure 13: Equivalent circuit representing a sample treated on both sides; R1 and R2 are the resistances of the coatings formed on both sides of the sample; C1 and C2 are the capacitors formed by the electrodes of the measuring system in contact with the surfaces on both sides of the sample and R3 represents the metal substrate.

This circuit consists of three resistors; R1 and R2 which represent the resistances of the ceramic dielectric coatings formed on the two surfaces of the substrate and R3 which represents the conductive metal substrate, and two parallel plate capacitors C1 and C2, formed by the dielectrics with resistances R1 and R2 between two plates: one of the electrodes of the impedance spectrometer and the metal substrate between the two coated faces.

To validate this equivalent circuit, impedance measurements were carried out on the samples treated at various times by PEO. Figure 14 shows the impedance spectrum obtained with a sample treated for 360 s and the corresponding simulated spectrum considering the equivalent circuit in Figure 25. As can be seen, there is excellent agreement between the measured and simulated values over the entire range of measured frequencies. The values obtained for the circuit components are shown in Table 2. The adjustments confirmed the expectation of obtaining high electrical resistance values in the ceramic coatings (R1 and R2) and low values in the metal substrate (R3).

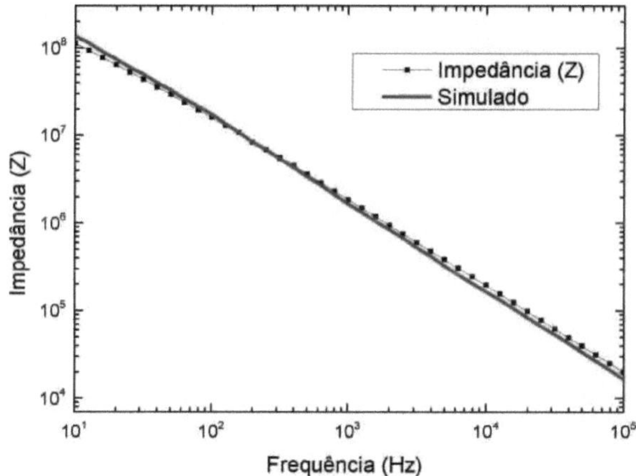

Figure 14: Measured impedance spectrum (dots) with a sample treated with for 360 s and the corresponding simulated spectrum (line) considering a circuit as shown in Figure 25.

Table 2: Component values obtained by fitting the equivalent circuit to the impedance spectrum for a sample treated for 360 s

Components	Measures
R1	2.66 x 107 Ω
R2	3.09 x 107 Ω
R3	5,81 Ω
C1	3.99×10^{-10} F
C2	2.20×10^{-10} F

Similarly, the respective components were obtained for the samples treated in duplicate, with treatment times ranging from 60 to 540 s. The results are shown in Table 3 for the average values of just one side of the sample.

Table 3: Electrical resistance values of coatings as a function treatment time

Time (s)	Electrical resistance (Ω)
60	19400
90	2,97E+08
120	4,12E+08
180	1,82E+08
360	4,03E+07
540	6,11E+07

With this, it can be seen that the electrical resistance increased abruptly until 120 s and then remained practically constant at around $\sim 5x10^7$ Ω until 540 s of treatment. The increase in electrical resistance as a function of treatment time is explained by the fact that the coating has become thicker, as shown in Figure 7, and by the increase in the proportion of aluminum oxide (Figure 12).

An important electrical property of the material is electrical resistivity, also known as specific resistance. The electrical resistances of coatings were shown above, but these are relative to the dimensions (thickness and area) of the sample. Electrical resistivity, on the other hand, is an inherent characteristic of the material and is independent of the dimensions of the sample (CALLISTER, 2007).

With the thickness and electrical resistance values, the electrical resistivity (specific resistance of the material) of the PEO coatings was calculated using the expression:

$$R = \rho \frac{l}{A} \tag{8}$$

Where ρ is the electrical resistivity (Ω-m), R is the electrical resistance (Ω), A is the area (m^2) and l is the thickness of the coating (m). The results are shown in figure 15.

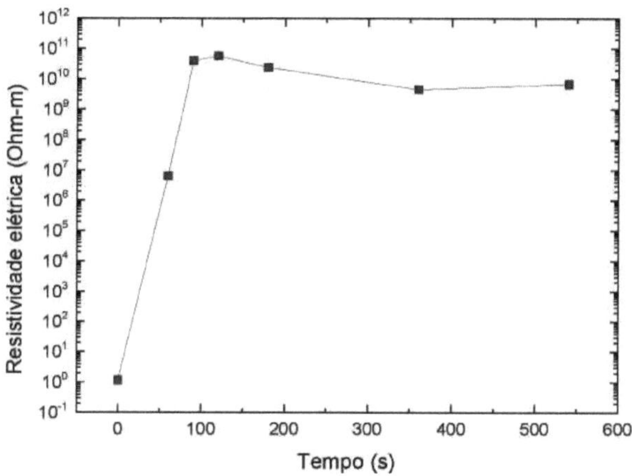

Figure 15: Electrical resistivity as a function of PEO treatment time. The value for 0 s was obtained with an untreated sample,

The sample treated with 60 s showed a relatively low electrical resistivity value (6.4×10^6), due to the fact that the coating was growing (initial oxidation phase), as shown by the EDS results, and therefore the substrate still influenced the characteristics of the coating. The maximum electrical resistivity of the coating reached 5.7×10^{10} Ω-m at 120 s of treatment and then oscillated between 2.4 x $^{10(10)}$ and 6.8 x $_{109}$ Ω-m for treatment times of 180 to 540 s. The high electrical resistivity observed in the coating is due to the formation of aluminum oxide, whose resistivity is ~10^{14} Ω-m (AUERKARI, 1996).

With 120 s of treatment with PEO using the parameters described in this work, it was possible to produce a ceramic coating with an electrical resistivity~ 10^{10}

times greater than untreated aluminum.

The capacitance of the coating was determined at an average frequency of 1000 Hz. Figure 16 shows the results as a function of treatment time.

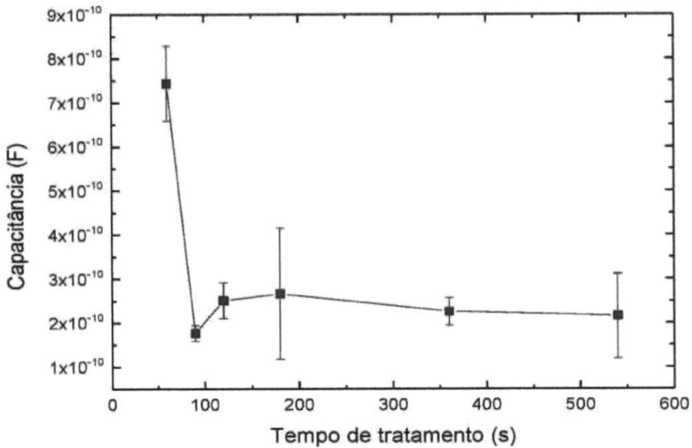

Figure 16: Capacitance of coatings as a function of PEO treatment time

The capacitance of the PEO-treated coatings varied little between the treatment times of 90 and 540 s, ranging from 1.76 x 10^{-10} to 2.67 x $10^{(-10)}$ F, with the exception of the coating on the sample treated with 60 s, which had a higher capacitance (7.43 x $10^{(-10)}$ F). After determining the resistance and capacitance values, equation 4 was used to calculate the impedances of the coatings in the frequency range between 10 and 100,000 Hz. Figure 17 shows the results obtained for treatments between 60 and 540 s.

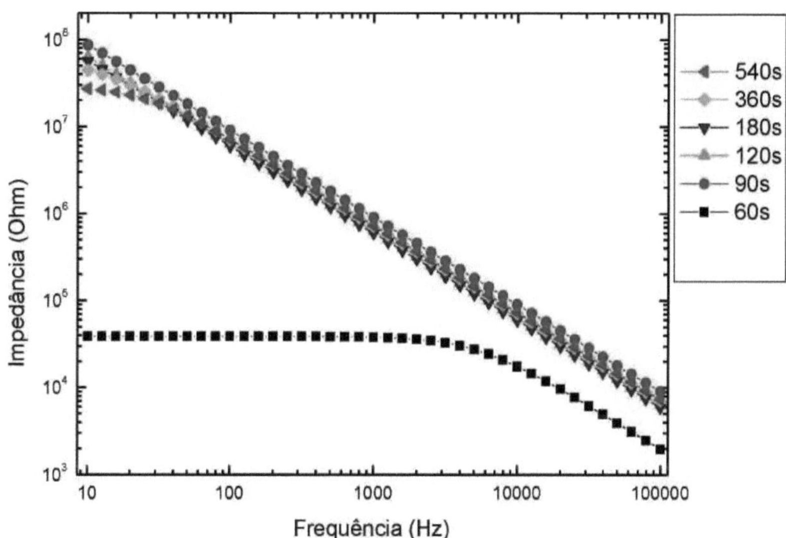

Figure 17: Impedance as a function of frequency for coatings obtained at various PEO treatment times.

For all PEO coatings at treatment times of 60 to 540 s, a decrease in impedance was observed as the frequency increased. This can be explained by the fact that the higher the frequency, the lower the capacitive reactance (*Xc*), as shown in equation 5, and since the impedance of this circuit is made up only of resistance and capacitive reactance, this determines the variation in impedance as a function of frequency.

The sample treated with 90 s had the highest impedance at all frequencies, reaching a maximum impedance value of 8.64×10^{7} Ω at 10 Hz. Although this sample did not have the highest resistance (Table 6), its capacitance was the highest of all the samples investigated (Figure 16). The coating obtained with 60 s of PEO resulted in the lowest impedance at all frequencies. This is due to the fact that this coating had the lowest resistance and the highest capacitance.

For a better understanding of the contributions of resistances and capacitances in the formation of impedance, let's analyze the coating treated with 540 s. This coating had the second lowest resistance (Table 6) and the second lowest capacitance (Figure 16) in the series, resulting in the second lowest impedance at low frequency and the second highest impedance at high frequency. This shows that for a resistor and capacitor circuit in parallel at low frequencies, the impedance will be determined by the resistance and at high frequencies it will be determined by the capacitance according to equation 4 (CHINAGLIA, 2008). This phenomenon occurs because in a parallel circuit, with a resistor and a capacitor at low frequencies, the flow of electrons will find it difficult to pass through the capacitor, due to its plate being full of electrons. In this way, the electric current will find it less difficult to flow through the resistor. And at high frequencies, as there is no time for electrons to accumulate on the capacitor plate and the resistor offers a certain resistance, electrons flow in greater quantities through the capacitor.

5 CONCLUSIONS

Ceramic coatings composed of crystalline aluminum oxides and amorphous silicon oxide were produced on the PEO-assisted aluminum substrate. According to literature data, the formation of crystalline aluminium oxide was expected due to oxidation of the aluminium surface and the incorporation of silicon into the coating comes from the electrolytic sodium silicate solution used in PEO.

As found in the literature, the thickness of these coatings grew with treatment time, reaching a maximum thickness of 2.83 μm at 540 s of treatment. A relatively high growth rate (over 1 μm/min) was observed in the first 90 s of treatment and, as the electrical resistance of the substrate increased, the growth rate decreased after 90 s of treatment.

In just 120 s of treatment, the coating obtained the highest electrical resistivity measurement (5.72×10^{10}). The high electrical resistivity observed in the coating is due to the formation of aluminium oxide, whose resistivity is ~10[(14)] Ω-m.

This treatment of the aluminium with PEO produced a ceramic coating with an electrical resistivity value 10^{10} times greater than that of the incoming aluminium.

REFERENCES

ANTÔNIO, C.A. **Electrolytic Plasma Film Deposition on Aluminum Alloys.** 2010. 96f. Dissertation (Master's Degree in Materials Science and Technology) - UNESP, Faculty of Sciences, Bauru, 2011.

ASM HANDBOOK, vol 2, Properties and selection: Nonferrous Alloys and Special-Purpose Materials. ASM International, 1992.

ASM HANDBOOK, vol 5, Surface Engineering. ASM International, 1994.

BRAZILIAN ALUMINUM ASSOCIATION. **Technical guide to aluminium - Surface Treatment.** 2nd ed. 2004.

ASM HANDBOOK, vol 5, Surface Engineering. ASM International, 1992

ATKINS, P.; JONES, L. **Principles of chemistry: questioning modern life and the environment.** Porto Alegre: Bookman, 2001, Chap. 2.

BRAZILIAN ALUMINUM ASSOCIATION. Guia tècnico do aluminio - Laminaçâo. 2° ed. 2004 apud ANTÔNIO, C.A. **Deposiçao de Filmes por Plasma Eletrolitico em Alizas de Aluminio**. 2010. 96f. Dissertation (Master's Degree in Materials Science and Technology) - UNESP, Faculty of Sciences, Bauru, 2011.

AUERKARI, P. **Mechanical and physical properties of engineering alumina ceramics.** Vuorimiehentie: VTT Manufacturing Technology, 1996. 11-13 p. ISBN 951-38-4987-2.

BOSTA, M, M, S, A.; MA, K, G.; CHIEN, H, H. The effect of MAO processing time on surface properties and low temperature infrared emissivity of ceramic coating on aluminium 6061alloy. **Infrared Physics & Technology**, v 60, p. 323-324, 2013.

BRAND, J.; GILS, S. V.; BEENTJES, P. C. J.; TERRYN, H.; WIT, J. H. W. Ageing of aluminum oxide surfaces and their subsequent reactivity towards bonding with organic functional groups. **Applied Surface Science,** v 235, p. 465-474, 2004.

CALLISTER, W. D. **Materials science and engineering: An introduction.** LTC: Rio de Janeiro, 2007, Chap. 3.

CHINAGLIA, D.L et al. Impedance spectroscopy in the teaching laboratory. **Revista Brasileira de Ensino de Fisica**, v. 30, p. 4054_1- 4054_9, 2008.

WEB COLLEGE. Available at< http://www.colegioweb.com.br/trabalhos-escolares/quimica/eletroquimica-i-pilhas/tabela-de-potencial-padrao-de-reduçãoo.html> Accessed July 2015.

Foucault's currents. Available at:< http://efisica.if.usp.br/eletricidade/basico/correntes_foucault/> Accessed July 2015.

DEHNAVI, V.; LUAN, B. L.; SHOESMITH, D. W.; LIU, X. Y.; ROHANI, S. Effect of duty cycle and applied current frequency on plasma electrolytic oxidation (PEO) coating growth behavior. **Surface & Coatings Technology**, v 226, p. 100-107, 2013.

DUNLEAVY, C.S.; GOLOSNOY, I. O.; CURRAN, J. A.; CLYNE, T. W. Characterization of discharge events during plasma electrolytic oxidation. **Surface and Coating Technology**, v.203, p.3410-3419, 2009.

GUPTA, P.; TENHUNDFELD. G.; DAIGLE, E. O.; RYABKOV, D. Electrolytic plasma technology: Science and Engineering - An overview. **Surface and Coating Technology**, v.201, p.8746-8760, 2007.

HALLIDAY, D.; RESNICK, R. **Fundamentals of Physics.** Ed. 8, Rio de Janeiro: Livros Técnicos e Cientificos Editora S. A, 2009, Chap. 3.

HAANAPPEL, V. A. C.; CORBACH, H. D. V.; FRANSEN, T.; GELLINGS, T. J. Properties of alumina films prepared by low-pressure metal-organic chemical vapor deposition. **Surface and Coatings Technology,** v 72, p. 13-22, 1995.

HUSSIEN, R.O.; NIE, X.; NORTHWOOD, D. O. An investigation of ceramic coating growth mechanisms in plasmaelectrolytic oxidation (PEO) processing. **Electrochimica Acta**, v.112, p. 111-1119, 2013.

INFOESCOLA. Available at:< http://www.infoescola.com/fisica/espectro-visivel/> Accessed July 2015.

JIANG, Y.; ZHANG, Y.; BAO, Y.; YANG, KE. Sliding wear behavior of plasma electrolytic oxidation coating on pure aluminum. **Wear**, v.271, p. 1667-1670, 2011.

KHAN, R.H.U.; YEROKHIN, A.; LI, X.; DONG, H.; MATTHEWS, A. Surface characterization of DC plasma electrolytic oxidation treated 6082 aluminium alloy: Effect of current density and electrolyte concentration. **Surface Coating Technology**, v.205, p. 1679-1688, 2010.

KRAPF, 2003 apud ANTÔNIO, C.A. **Deposition of Films by Electrolytic Plasma on Aluminum Alloys.** 2010. 96f. Dissertation (Master's Degree in Materials Science and Technology) - UNESP, Faculty of Sciences, Bauru, 2011.

LI, Q.; LIANG, J.; LIU, B.; PENG, Z.; WANG, Q. Effects of cathodic voltages on structure and wear resistance ofplasma electrolytic oxidation coatings formed on aluminium alloy. **Applied Surface Science**, v 297, p. 176-181, 2014.

LIF, J.; ODENBRAND, I.; SKOGLUNDH, M. Sintering of alumina- supported nickel particles under amination conditions: Support effects. **Applied Catalysis A: General,** v 317, p. 62-69, 2007.

LISBÔA, A, J, T. et al. Study of the thermal resistance of alpha-phase aluminum oxide (α-Al2O3) films deposited on the paper substrate. **Ceramics International**, v 40, p. 9509-9516, 2014.

LUGOVSKOY, A.; ZINIGRAD, M.; KOSSENKO, A.; KAZANSKI, B. Production of ceramic layers on aluminum alloys by plasma electrolytic oxidation in alkaline silicate electrolytes. **Applied Surface Science**, v.264, p. 743-747, 2013.

MACDONALD, J.R; BARSOUKOV, E. **Impedance Spectroscopy Theory,**

Experiment, and Applications. New Jersey: John Wiley & Sons, Inc., 2005.

MARTIN, J. et al. Effects of electrical parameters on plasma electrolytic oxidation of aluminum. **Surface Coating Technology**, v.221, p. 70-76, 2013.

MOHANNAD, M. S. A. B.; MA, K. J.; CHIEN, H. H. The effect of MAO processing time on surface properties and low temperature infrared emissivity of ceramic coating on aluminum 6061 alloy. **Infrared Physics & Technology**, v.60, p. 323-334, 2013.

MONTERO, I; FERNANDEZ, M; ALBELLA, J. M. Pore formation during breakdown process in anodic Ta2O5 films. **Electrochimica Acta**, v. 32, n. 1, p. 171-174, 1987.

OLIVEIRA, C.R. **Alteration of the Surface Properties of Aluminum via Plasma Electrolysis.** 2009. 122f. Dissertation (Master in Materials Science and Technology) - UNESP, Sorocaba, 2009

PADILHA, A.F. **Materiais de engenharia: microestrutura e propriedades.** Curitiba: Hemus livraria, distribuidora e editora S. A, 2000, Chap. 6.

PAULMIER, T.; BELL, J. M.; FREDERICS, P. M. Development of a novel cathodic plasma/electrolytic deposition technique part 1: Production of titanium dioxide coatings. **Surface and Coatings Technology**, v 201, p. 8761-8770, 2007.

PARFENOV, E.V.; YEROKIN, A.L.; MATTHEWS, A. Impedance spectroscopy characterization of PEO process and coatings on aluminium. **Thin solid films**, v.516, p. 428-432, 2007.

PARFENOV, E.V.; YEROKIN, A.L.; MATTHEWS, A. Frequency response studies for the plasma electrolytic oxidation process. **Surface Coating Technology**, v.201, p. 8861-8670, 2007.

QNix® 7500: The 'junior Jack' of all trades. Available at: < http://www.easyfairs.com/uploads/tx_ef/7500_englisch-d426dd.pdf>.

Accessed July 2015.

RUSSEL, J.B. **General Chemistry.** Ed. 2, Sâo Paulo: Makron Books, 1994.

SKOOG, D.A.; HOLLER F.J.; NIEMAN T.A. **Principles of instrumental analysis.** 5 ed., Porto Alegre, Bookman,2002, Chap. 12,16,17.

TIAN, J; LUO, Z; QI, S; SUN, X. Structure and antiwear behavior of microarc oxidized coatings on aluminum alloy. **Surf. Coat. Technol**., n. 154, p. 1-7, 2002.

YASUDA, H.J. **Plasma Polymerization.** Academic Press. Inc, New York, 1985.

YEROKHIN, A.L. et al. Oxide ceramic coatings on aluminum alloys produced by a pulsed bipolar plasma electrolytic oxidation process. **Surface Coatings Technology**, v.199, p.150-157 2005.

YEROKHIN, A.L.; NIE, X.; LEYLAND, A.; MATTHEWS, A.; DOWEY, S. J. Plasma electrolysis for surface engineering. **Surface Coating Technology**, v.122, p. 73-93, 1999.

YEROKHIN, A.L.; LYUBIMOV, V. V.; ASHITKOV, R. V. Phase formation in ceramic coatings during plasma electrolytic oxidation of aluminum alloys. **Ceramics International**, v.24, p. 1-6, 1998.

WEI-CHAO, G., et al. 2007 apud ANTÔNIO, C.A. **Electrolytic Plasma Film Deposition on Aluminum Alloys**. 2010. 96f. Dissertation (Master's Degree in Materials Science and Technology) - UNESP, Faculty of Sciences, Bauru, 2011.

WEI, T; YAN, F; TIAN, J. Characterization and wear- and corrosionresistance of microarc oxidation ceramic coatings on aluminum alloy. **Journal of Alloys and Compounds**, v. 389, p. 169-176, 2005.

Printed by Books on Demand GmbH, Norderstedt / Germany